# AMMONIUM NITRATE SPECIFICATIONS
# &
# QUALITY OF EXPLOSIVES FOR ROCK BLASTING

## MANOJ KUMAR PATEL
M.B.A., Ph.D.

IPUL, ROURKELA, INDIA

This book has been published with all efforts taken to make the material error-free. This is a technical book containing materials, and data for practical uses by professionals in the fields of ammonium nitrate, explosives (for rock blasting), and mining.

Dr. Manoj Kumar Patel
IPUL, Rourkela, India

# THE GIST

This book is a ready reckoner and handy guide to the purchasers of ammonium nitrate and manufacturers of explosives (containing ammonium nitrate).

It is written to facilitate understanding the nature of technical specifications in finalizing purchases of ammonium nitrate.

The contents will help the purchasers in emphasizing and asking for the correct parameters of AN from the AN sellers.

# INDEX

# CHAPTER I

# SPECIFICATION

A specification refers to a set of documented requirements to be satisfied by a material. It is a technical standard.

The specification is important to understand the quality of a material. Customers review the specifications to decide the purchase.

There are different types of technical, engineering, product, process, and services specifications, and the term is used differently in different technical contexts.

Many product development methods describe product development as a well-structured, sequential flow starting with a requirements specification, or design specification, and ending with a product solution.

Thus establishing a design specification is essential and should be a central issue in design research. Several procedures for creating requirements for a specification are suggested in literature acknowledging the benefit of having a comprehensive specification at the start of the design process. Furthermore, these works of literature provide some guidelines for the preparation of a good specification i.e. requirements should be unambiguous, solution-independent formulation, clearly linked to customer needs, measurable (qualitative or quantitative).

To ensure the specification is more practical, several authors have distinguished requirements into demand and wishes. However, the contents and form of a specification are different from case to case and are influenced by several factors: the complexity of the factions; complete plant vs. machine component, design difficulty; new design and development vs. adaptive design, the requirement for additional properties; safety, life, appearance, and problem initiator or sponsor.

To identify a complete set of requirements for a product at the early phase of the design process becomes essential yet impossible in reality. Based on an empirical study, - completeness is a criterion that is basically unachievable. Often new requirements are included as the design process proceeds into the advance state because we do not know the problem fully until the solution is created. The requirements are changed (i.e.

developed, explored and expanded) during the design process into a more complete description or final specification.

In high energy materials like ANFO, Site mix emulsion (SME), Site mix slurry (SMS), package emulsion, and package slurries, the major component is ammonium nitrate. It constitutes about 60 – 95% in the products of different types. So, specifications of AN are important to achieve good quality of high energy materials.

Manufacturers of ammonium nitrate provide technical specifications to the purchasers and users. Some manufacturers provide detail specification. Their specification sheets contain as many as 25 different quality parameters. Some other manufacturers, on the other hand, supply specifications containing as less as four parameters. Thus sometimes it becomes difficult to understand the quality of the ammonium nitrate for specific uses like ANFO, SME, SMS, and so on.

This book is written to serve as a ready reckoner and handy guide to the manufacturers of explosives (containing ammonium nitrate).

The book aims to understand the nature of specification finalization during the purchase of ammonium nitrate for use. To achieve these aims the chapters in this book have been written to understand (a) the parameters that influence the development of a good specification in product development in a collaboration project between a manufacturer and the purchaser, (b) the AN manufacturer (seller) define the content of a specification, and (c) mode of specifications used during the design process.

Let us see it for ammonium nitrate. In the next chapter, we have started with a specification of AN having 22 quality parameters. While these parameters are important it is equally important to prune it to fewer parameters that will define all other parameters relevant to explosive manufacture. AN specification is reviewed for parameters and has been made concise, specific, and useful.

*mkp*

# CHAPTER II

# AMMONIUM NITRATE

## INTRODUCTION

Ammonium nitrate is one of the most preferred oxidizers in the manufacture of improvised explosives like watergel, slurry, emulsion, and ANFO. An effective oxidizer can simply consist of ammonium nitrate alone. The fuels could be many, such as sulfur, urea, coarse grade metal powder, diesel oil, furnace oil, LSHS, waxes, and so on.

Thus ammonium nitrate is near-indispensable in explosive for rock blasting. It is equally important that ammonium nitrate should satisfy quality parameters for its remaining effective.

Bodies of ammonium nitrate mean, but are not limited to; formulations of ammonium nitrate, physical forms, such as prills, granules, or said prills and/or granules in an emulsion, and compositions wherein ammonium nitrate is a major component. There are other properties and presence (rather absence) of other chemicals (even in traces) important for its use in explosives.

The qualities of AN for watergel, bulk SME, cartridge emulsion, and ANFO have to be different to achieve the best quality of the respective products.

| No. | PARAMETER | SPECIFICATION of AN for | |
| --- | --- | --- | --- |
| | | Watergels and Emulsion | ANFO |
| 1 | Appearance | Without any visual impurities | Without any visual impurities |
| 2 | Physical form | Free flowing, spherical prills | Free flowing, spherical prills |
| 3 | Purity | 99.0% Minimum, on dry basis | 98.0% Minimum, on dry basis |
| 4 | friability (strength) | | |

| 5 | Shelflife of prills in the form of prills | 3 months | 6 months |
|---|---|---|---|
| 6 | Purity | 34.65% Minimum | 98.0% Minimum, on dry basis |
| 7 | Nitrogen content | 34.65% Minimum | 34.30% Minimum |
| 8 | Moisture | 0.5% Maximum | 0.5% Maximum |
| 9 | pH of 10% Solution | 5.0 to 6.0 | 5.0 to 6.0 |
| 10 | Phosphates | Nil | Not Applicable |
| 11 | Calcium as $Ca(NO_3)_2$ | Nil | Not Applicable |
| 12 | Magnesium as $Mg(NO_3)_2$ | Nil | Not Applicable |
| 13 | Sulphates | Nil | Not Applicable |
| 14 | Chlorides | Nil | Not Applicable |
| 15 | Nitrite as Ammonium Nitrite | 0.02% maximum | Not Applicable |
| 16 | Turbidity | Nil | Not Applicable |
| 17 | Bulk density | 1.00 to 1.05 g/cc | 0.70 to 0.80 g/cc |
| 18 | Water Insoluble Matter | 0.1% Maximum | 0.1% Maximum |
| 19 | Particle size<br>More than 3 mm<br>1 to 3 mm<br>Less than 1 mm | 3% Maximum<br>95% Minimum<br>2% Maximum | Nil<br>98% Minimum<br>2% Maximum |

| | | | |
|---|---|---|---|
| 20 | Lab Batch to be Made for Gel Stability | Gel should retain it characteristics for 8 weeks in Hot Storage Test at 45 – 50OC | Not Applicable |
| 21 | Oil Absorption | Not Applicable | 6.0 - 6.5 % |
| 22 | Oil Retention (up to 72 hours) | Not Applicable | 5.5 % minimum |

Purity remaining the same, the presence (or absence) of other physical and chemical parameters make differences in accepting ammonium nitrate for specific uses. Say, for example, the organic coating material makes AN the best choice for use in the manufacturing of ANFO, but makes the same material the worst choice for watergel, slurry, and emulsion explosives.

Understanding the role and importance of each parameter in the specification sheet is of importance to produce explosives of the best quality and shelf life.

## PURITY

A molecule of ammonium nitrate contains – two atoms of nitrogen, four atoms of hydrogen, and three atoms of oxygen. So from the viewpoints of Stoichiometry, atomic weight, and molecular formula, it confirms that the presence (experimental determination through assay) of 35% of nitrogen is equivalent to 100 percent purity of ammonium nitrate. Commercially available AN, however, comes with the presence of foreign materials in the form of physical materials like dust, and chemical materials like salts of magnesium, calcium, chloride, etc. Thus expecting 35% nitrogen in commercial-grade AN is impractical.

On average, the nitrogen content in commercial-grade AN varies in the range of 98 – 99%. The nitrogen content is dependent on source materials, technology adopted, and handling and storage of input and finished materials in different stages like transportation and storage. Manufacturers consider these to mention the purity (back-calculated from the nitrogen content obtained on the chemical analysis) of their material.

As a rule of thumb, purity in the range of 98 – 99 percent should be alright for explosive manufacturing. But it is worth remembering that purity alone is no guarantee for the release of gaseous and heat energy when used in explosive for rock blasting. A consignment of AN having 98% purity may yield better results compared to another consignment of AN with a purity of 99 percent.

## FRIABLIITY

Friability is an important parameter but often ignored or gets camouflaged by other parameters like purity, augurability, and agglomeration.

Friability is the percentage of weight loss of powder from the surface of the granules (prills) due to mechanical action during transportation and auguring. Granules and prills tend to powder, chip, and fragment.

In simple words, the friability test tells how much mechanical stress the AN granules are able to withstand during their manufacturing, distribution, and handling by the customer.

It also plays a very important role in its marketing and dissolution (and thus productivity).

A friability value provides a measure of granule's weakness. It is correlated with the crushing strength of the granules (prills). There is an increase in crushing strength with a corresponding decrease in friability values with binder concentration for all formulations.

$$\text{Friability} \propto (1 \:/\: \text{crushing strength})$$

Friability should be considered an important parameter for porous prilled ammonium nitrate (PPAN). However, this parameter may be discounted for high-density ammonium nitrate (HDAN). Because (a) PPAN is suitable for ANFO manufacturing for its porosity and lower density and (b) HDAN is used for the manufacture of watergel, and slurry explosives for its ease for solubility in an aqueous medium to prepare the oxidizer blend.

A problem in the ammonium nitrate art is the competing need for porosity versus the need for hardness. This is especially true for ammonium nitrate bodies that use explosive applications.

Hardness is important for AN for use in watergels and packaged emulsion products (since hardness is related to the leaching rate of the hard crystals into the aqueous medium (OB). Here the AN is of density 1.00 – 1.05 g/cc.

Porosity is important for use in ANFO (where oil has to leach into the prills and get encapsulated inside the hollow spheres of porous AN prills). Here the AN is of low bulk density 0.70 – 0.80 g/cc.

## Impact of friability parameter

In ANFO it is required that the prills remain intact in their prill form so that these can absorb and retain oil. If friability is inadequate then the broken prills will not give rise to ANFO. The oxygen balance will get offset. Energy release from the blast will be lesser. At the same time, both orange (due to the presence of broken AN, and insufficient or unabsorbed oil) and black (due to fuels available free and unabsorbed) will emanate post-blast in the mines. Thus improper friability percentage will cause negative results in (a) energy release, (b) incomplete combustion, and (c) emission of toxic gases of oxides of nitrogen and carbon.

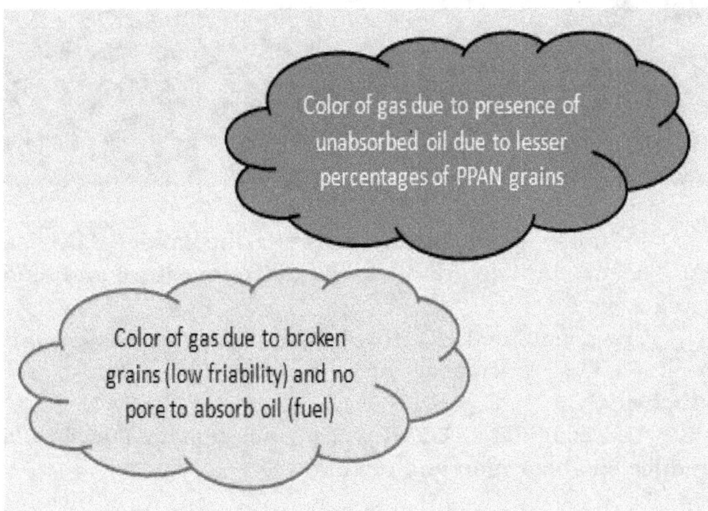

Friability is important for both the grades LDAN and HDAN because broken granules will attract moisture and tend to coagulate.

The coagulated HDAN will take higher time to get mixed into the OB (increase fuel consumption, and reduce productivity).

The coagulated LDAN will (a) not remain flowable, and (b) reduce oil absorption and retention.

## Testing for friability

Friability testing is a laboratory technique often used by the pharmaceutical industry involves repeatedly dropping a sample of tablets over a fixed time, using a rotating wheel with a baffle.

$$Friability = [(W1 - W2) / W1] \times 100$$

Where,
$W_1$ = Weight of granules (Before Tumbling) &
$W_2$ = Weight of granules after Tumbling or friability)

*Limit: Friability (%) = Not More Than 1.0 %*

Generally, the test is run for once. If any cracked, cleaved or broken granules present in the sample after tumbling, the consignment fails the test.

Hardness testing is a laboratory technique used by the pharmaceutical industry to determine the breaking point and structural integrity of a tablet and find out how it changes "under conditions of storage, transportation, packaging and handling before usage" The breaking point of a tablet is based on its shape.

Hardness testing machines perform three common kinds of scientific hardness tests such as the Brinell hardness, the Rockwell hardness, and the Vickers hardness.

However, one can always use a conversion table for comparing the Rockwell (B & C), Vickers, and Brinell values. Tables like these are not 100% accurate but give a good indication.

| Brinell Hardness | Rockwell | Rockwell | Vickers | N/mm² |
|---|---|---|---|---|
| HB | HRC | HRB | HV | |
| 469 | 50 | 117 | 505 | |
| 468 | 49 | 117 | 497 | |
| 456 | 48 | 116 | 490 | 1569 |
| 445 | 47 | 115 | 474 | 1520 |
| 430 | 46 | 115 | 458 | 1471 |
| 419 | 45 | 114 | 448 | 1447 |

| 415 | 44 | 114 | 438 | 1422 |
|---|---|---|---|---|
| 402 | 43 | 114 | 424 | 1390 |
| 388 | 42 | 113 | 406 | 1363 |
| 375 | 41 | 112 | 393 | 1314 |
| 373 | 40 | 111 | 388 | 1265 |
| 360 | 39 | 111 | 376 | 1236 |
| 348 | 38 | 110 | 361 | 1187 |
| 341 | 37 | 109 | 351 | 1157 |
| 331 | 36 | 109 | 342 | 1118 |
| 322 | 35 | 108 | 332 | 1089 |
| 314 | 34 | 108 | 320 | 1049 |
| 308 | 33 | 107 | 311 | 1035 |
| 300 | 32 | 107 | 303 | 1020 |
| 290 | 31 | 106 | 292 | 990 |
| 277 | 30 | 105 | 285 | 971 |
| 271 | 29 | 104 | 277 | 941 |
| 264 | 28 | 103 | 271 | 892 |
| 262 | 27 | 103 | 262 | 880 |
| 255 | 26 | 102 | 258 | 870 |
| 250 | 25 | 101 | 255 | 853 |
| 245 | 24 | 100 | 252 | 838 |
| 240 | 23 | 100 | 247 | 824 |
| 233 | 22 | 99 | 241 | 794 |
| 229 | 21 | 98 | 235 | 775 |
| 223 | 20 | 97 | 227 | 755 |
| 216 | 19 | 96 | 222 | 716 |
| 212 | 18 | 95 | 218 | 706 |
| 208 | 17 | 95 | 210 | 696 |
| 203 | 16 | 94 | 201 | 680 |
| 199 | 15 | 93 | 199 | 667 |
| 191 | 14 | 92 | 197 | 657 |
| 190 | 13 | 92 | 186 | 648 |
| 186 | 12 | 91 | 184 | 637 |
| 183 | 11 | 90 | 183 | 617 |
| 180 | 10 | 89 | 180 | 608 |
| 175 | 9 | 88 | 178 | 685 |
| 170 | 7 | 87 | 175 | 559 |
| 167 | 6 | 86 | 172 | 555 |

| 166 | 5 | 86 | 168 | 549 |
|-----|---|----|-----|-----|
| 163 | 4 | 85 | 162 | 539 |
| 160 | 3 | 84 | 160 | 535 |
| 156 | 2 | 83 | 158 | 530 |
| 154 | 1 | 82 | 152 | 515 |
| 149 |   | 81 | 149 | 500 |

The SI unit of hardness is $N/mm^2$. The unit Pascal is thus used for hardness as well but hardness must not be confused with pressure.

## Maintaining friability

The maintenance of quality of granular and porous AN in blasting applications depends on variables like, the technology, the applicator, the operators, the material quality, the weather condition, the field. These are:

1. Weather = precipitation, Humidity, Wind
2. Material properties = Physical, and Chemical
3. Operator = Education, Awareness, Aptitude
4. Applicator = Setup, Maintenance, Calibration
5. Technology = Setup, Firmware, data
6. Field = Water, Moisture, Depth, Sleep time

The 5Rs of friability in the blasting stewardship include Right source, Right transportation, Right storage, Right types of machinery (BMD vehicle), and Right use

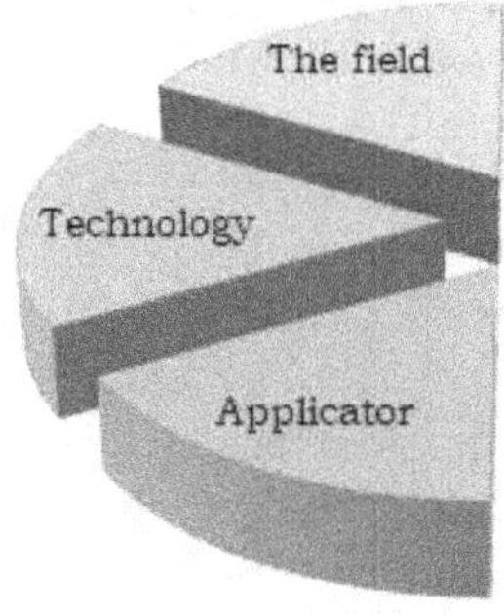
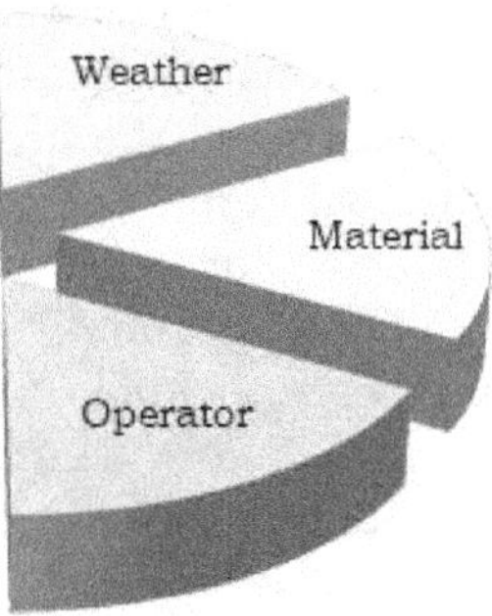

Some of the important aspects that give indication about friability of prill porous AN (PPAN) suitable for ANFO is:

⇒ Uniqueness of the physical properties from crystalline and solid AN.
⇒ Size and uniformity of particle size
⇒ Particle density
⇒ Bulk density
⇒ Flow characteristics
⇒ Coefficient of friction
⇒ Anti caking character
⇒ Oil absorption and retention capacity

## SHELFLIFE OF PRILLS

Purity remaining intact, damage of prills into powder or formation of prills into agglomerated mass is the end of shelflife. These are no more useful for ANFO manufacturing. In the case of uncoated material, these pose difficulties in handling, transportation, and use.

Ammonium nitrate surface is extremely hygroscopic, due to high energy on its surface and increases the potential of the molecules to absorb moisture from the environment. Absorption of moisture leads to oozing out of material. That is the loss of material on storage. Agglomerated masses on the other hand consume higher energy (steam) to get into the solution.

The Shelflife of prills is improvised by coating these with hydrophobic materials. These hydrophobic materials are termed with the common name of surfactants. The surfactant encapsulates the ammonium nitrate when the prilling process is going on from molten AN in the manufacturing plant.

Myriads of surfactants are available and used in coating the AN prills to protect the prills from the absorption of moisture from the atmosphere. Some of these surfactants are; MgO, alcohol, and acid surfactant groups with hydrophobic tails from C14 to C22, cetyl alcohol or stearic acid or stearyl alcohol, acetonitrile, cetyl alcohol, stearyl alcohol, stearic acid, and myristic acid. These surfactant materials have a good performance of anti-hygroscopicity. Usually, there are two coatings given to the prills. In some cases, the first coating layer surfactant will be by stearic acid or cetyl alcohol, because they have a good performance of anti-hygroscopicity and less amount dissolve in the second coating solvent.

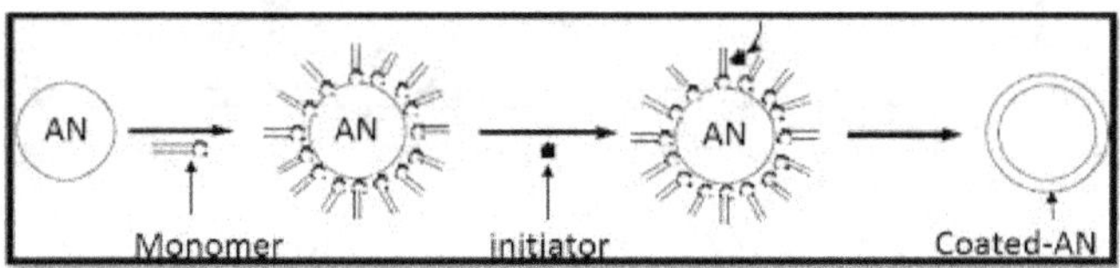


While coating layers enhance shelflife of prills the same materials pose operational problems and subsequently shelflife reduction of explosives which are manufactured using coated prills. These difficulties are noticed by visual observation of the aqueous solutions of AN prills (in a glass beaker or test tube) while making a quality check on turbidity (opacity) before using the same for explosives manufacturing.

For the users of AN it is necessary to find out the percentage of the coated material. This determination is carried out in many different methods including the one described herein. It is termed as – Mass Ratio Coating Layer (MRCL). This parameter is determined from the results of (a) moisture absorption ratio, and (b) decline in moisture absorption rate. Thus we shall cover this matter in the section under and (b) the moisture absorption rate.

Coated prills are not advisable for use in watergels, slurry, and water-in-oil emulsion explosives. Coated prills should however be preferred for the manufacturing of ANFO.

## TURBIDITY

Turbidity is a lack of transparency. So materials that do not get dissolved in water are the reasons for turbidity in aqueous solutions. Turbidity thus is the cloudiness or haziness of a fluid caused by large numbers of individual particles that are generally invisible to the naked eye, similar to smoke in the air. The measurement of turbidity is useful to assess the suitability of the consignment of ammonium nitrate for use in the manufacture of explosives. As a rule of thumb, a consignment of AN giving rise to a turbid solution is unsuitable for use in watergels, slurry, and emulsion explosives. The same material however may be used for the manufacture of ANFO.

Turbidity is measured in NTU (Nephelometric Turbidity Units). The simplest and lowest-cost way to measure the turbidity of a sample is a turbidity tube. This is a tube with a black cross at the bottom and the user simply keeps pouring water into the tube until they can no longer make out the black cross at which point you can read off the scale on the outside of the tube in NTU.

A Nephelo meter or turbidity meter also is used to measure turbidity quantitatively of AN solutions.

Turbidity of 10 NTU and more is detrimental to the shelf life of watergel, slurry, and emulsion explosives. The same turbidity is however not affecting the quality of ANFO explosives does.

## MOISTURE CONTENT

Moisture content in the prills is important from handling, storage, and usage points of view. This parameter is related to the friability, and the surfactant types and contents of the prills. Surfactants we know are the hydrophobic or anti-hygroscopic materials used to encapsulate the prills of AN. If these materials are used properly then moisture content on AN prills remains under control. Thus the consignments of AN also remain free-flowing or augurable.

From the important viewpoints of friability, the number and percentages of surfactants, and the capacity to hinder moisture absorption, it is necessary for the users of AN to assess (a) the moisture content, (b) the moisture absorption ratio, and (c) the mass ratio of the coating layer.

The moisture content (or the moisture absorption rate) can be measured by the simple method of determining the weights of a specific quantity of sample before and after drying (for 24 hours at 35 °C).

$$MAR = [(W1 - W2)/W1] \times 100 \qquad \ldots \qquad (1)$$

Where

W1 = weight of the sample before drying, and

W2 = weight of the sample after drying

The decline in the moisture absorption rate (MAR) is determined by carrying out the experiment for an uncoated sample of AN and then comparing it with the result obtained in eqn 1.

$$D_{MAR} = (MAR1 - MAR)/MAR1] \times 100 \qquad \dots \qquad (2)$$

Where
DMAR = decline in moisture content
MAR = figure from eqn 1.
MAR1 = MAR of uncoated sample

Now, continuing the thread from the SHELFLIFE section and then combining it with the previous two equations, the mass ratio of the coating layer (MRCL)is determined.

The method to determine MRCL:

Step 1: Wet a filter paper of say 18 cm diameter with deionized water.

Step 2: Keep the filter paper in an oven at 100 °C, and then take its weight (Wb).

Step 3: The sample of the coated AN after absorption humidity was dissolved in deionized water, and filtered through a tapered funnel, and the filter paper was washed several times with deionized water, and placed in an oven at 100 °C for 1 hour.

Step 4: The filter paper was weighed again ($W_a$).

The mass ratio of the coating layer was calculated by the following equation:

$$MRCL \, (\%) = [M1 \, / \, M0] \times 100 \qquad \dots \qquad (3)$$

Where,
M1 = (Wa – Wb), and
M0 = W1 (from eqn 1)

## PARTICLE SIZE

Particle size plays important role in LDAN and PPAN used for ANFO manufacturing. Uniformity in sizes helps in quantitative determinations of

⇒ Bulk density,
⇒ Oil absorption uniformity,

⇒ Flow rate of AN, and FO to deliver consistent quality of ANFO from the BMD vehicles.

⇒ Quantity of charges per meter of blast holes in the mines, and

⇒ Setting the flow meters of the BMD vehicle.

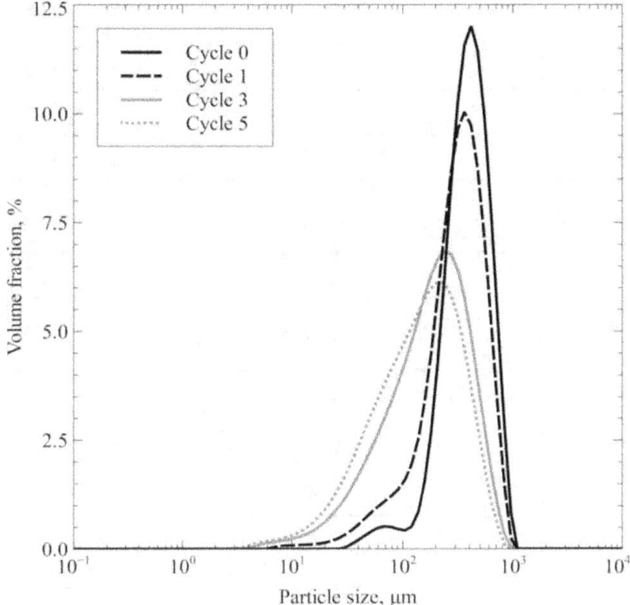


Particle size affects different activities. Smaller particle sizes increase the surface area and thus increase the sensitivity of AFO made out of these. Some close correlations between particle sizes and other parameters are:

1. Particle size and crystallite size
2. Particle size and bulk density relationship
3. Particle size and solubility
4. Particle size and size distribution
5. Particle size and shape
6. Particle size and dissolution rate.

Particle sizes affect cost aspects also to the users. a calculation showing the cost impact, both adverse and favourable, is given below:

| Sl. No. | Parameter | UoM | Values | |
|---|---|---|---|---|
| 1 | AN density | t/m³ | 0.75 | 0.8 |
| 2 | FO density | t/m³ | 0.81 | 0.81 |
| 3 | ANFO (94.5:5.5) density | t/m³ | 0.753 | 0.801 |
| 4 | AN Cost | Rs/t | 30000 | 30000 |
| 5 | FO Cost | Rs/t | 98765 | 98765 |
| 6 | ANFO Cost | Rs/t | 33782 | 33782 |
| 7 | Blast hole diameter | m | 0.311 | 0.311 |
| 8 | ANFO Cost | Rs/m | 1934 | 2055 |
| 9 | No of holes charged | number | 50 | 50 |
| 10 | Length of charged hole | m | 20 | 20 |
| 11 | Cost of ANFO | Million Rs/day | 1.93 | 2.06 |
| 12 | Savings | Million Rs/year | 44.28 | |

(Conversion between Rs and USD as on 14th October 2020, is Rs. 75 = USD 1).

Particle size, thus, is not only a quality influencer but also a cost influencer.

## WATER INSOLUBLE MATTER (WIM)

The presence of silica, dusts, and other materials reflect as water insoluble matters in the AN. These are different from soluble chemicals like chlorides, sulphates, nitrites of ammonium ion. These are also in addition to the surfactants used in coating to protect AN from attracting moisture (surfactants are anti caking and anti hygroscopic).

Water insoluble materials (WIM) usually do not pose problem to the quality of explosives or ANFO, but affect pumps especially with respect to wear and tears. WIM also affects the pumping efficiencies and flow consistencies of AN solutions. This is because the WIMs choke the strainers (sieves) adjacent to pumps.

## UNIFORMITY OF THE PARTICLES

Imagine a situation of a restaurant chain in a big city like Hyderabad. Say the chain has got 23 restaurants in the city. *Dum biriyani* is its top of the list food in the menus in the city.

About 23 tons of *dum biriyani* are required, and there are 7 kitchens to cook it. Assuming that each kitchen prepares identical quantity of *dum biriyani,* the distribution of rice, and chickens will be made accordingly. As is favoured by the customers, the variety of the rice is one, but the sizes and varieties of chicken pieces in *dum biriyani* are (a) wings, (b) breast, (c) with bones, and (d) boneless. The survey has shown that the distribution of rice and four different varieties of chicken pieces per platter is followed with the recipe of 175g, 50g, 40g, 40g, and 45g, respectively. There are 23 restaurants and together they need 15000 tons of *dum biriyani* every day to serve their 55000 customer per day. For ease of business and distribution the owner has centralized the cooking in three different places, say kitchen one, kitchen two, and kitchen three. Each kitchen thus cooks about 5000 tons of *dum biriyani.*

To achieve the customer delight, each plate should contain ingredients as per the recipe. If recipe is not evenly distributed in each plate then:

1. Quantity of the ingredients issued from the store will match the consumption. That means there will be nil material variance, which is excellent, but,
2. Each customer will find disproportionate quantity of the wings, breasts, bone meat, and boneless meat.

The customer satisfaction will reduce and business will move south in the graph.

Identical is the condition with AN for ANFO. If prill sizes vary too wide then (a) the material variance may remain excellent (i.e. zero), but blast performance will be different for different blast holes. The overall blast will be sub-standard.

Uniformity in prill sizes is thus important.

## OIL ABSORPTION AND RETENTION

Prills with pores will absorb oil. That is very simple and fundamental. The point is "oil retention". If absorbed oil is not retained within the pores then the oil will gradually ooze out. And the oozed out oil is a straight loss because it will burn but will not take part in the explosion (detonation) reaction. The purchaser of AN (PPAN and LDAN) should be extremely careful of this parameter, because

$$\text{Oil absorption} \neq \text{oil retention} \quad \ldots \quad (1)$$

Oil retention should be the parameter to accept or do otherwise with the material.

## AGLOMERATION

Humidity and high temperature are adverse conditions for AN prills. Individual prills tend to absorb moisture and agglomerate. Agglomerated AN become non-usable for ANFO BMD vehicles. Prills do not flow and blast hole loading becomes impossible.

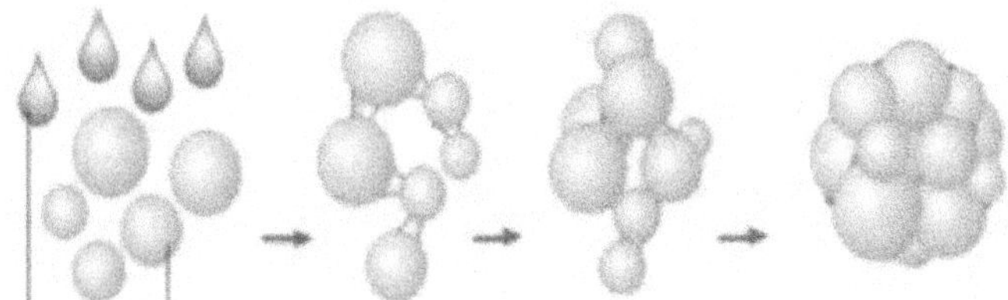

## CONTROL BATCH

Most of the specification sheets do not reflect this as quality parameter. But if implemented then this alone can replace all other quality parameters that we have discussed above. According to this, the user has to prepare control batches by conducting Design of Experiments (DoE) or better known as Taguchi's Design of Experiment. If the outcome of batches satisfy the quality parameters of ANFO, SMS, and SME, then the consignment of AN can be accepted with nil risk. (Taguchi's Design of Experiment is described in one of the books of the author).

The below shown is a photograph of a slurry watergel cap sensitive explosive. This product was made by the author as a control batch in one of the explosives manufacturing industries in India, in the month of September, 2019.

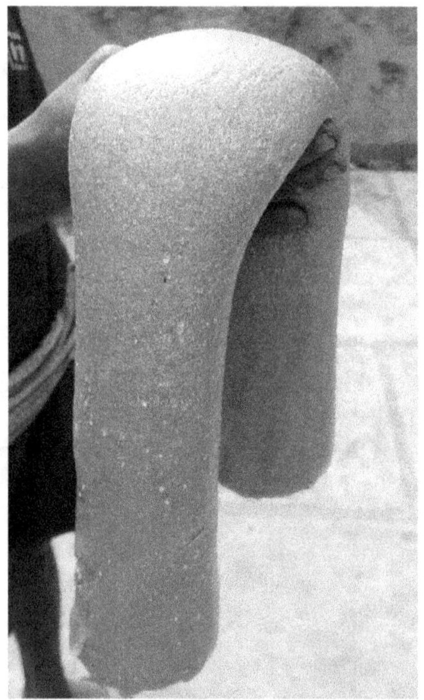

## BEFORE WE CONCLUDE

The purchaser sitting in the office and the user working in the mines often disagree on – relevance of specifications. The purchaser always retorts and defends him whenever problem encountered during use is mentioned by the user. The single most point counterpoint is:

User:            The supplied AN is not working like earlier materials.

Purchaser:    I have purchased the material as per the specifications.

Both of them are correct in their points but there is something amiss in correlating the parameters in the specification and the adverse result there from.

Case I:

Say the purchase of the AN is for making ANFO. The AN content mentioned in the specification sheet is 98% minimum. So, the supplier has supplied materials by sticking to the specification. Consider three different purchases. Then the effect on different output parameters that the blasting engineer in the mining witness specially in terms of colour of the fumes is – towards black smoke.

| AN | | | Yield from ANFO | | | | |
|---|---|---|---|---|---|---|---|
| AN Purity | actual AN in the ANFO | FO | OB | SF | AWS | ABS | VE |
| 100 | 94.5 | 5.5 | 0.09 | 18.81 | -917 | -734 | 375 |
| 99.75 | 94.3 | 5.5 | 0.05 | 18.81 | -917 | -734 | 374 |
| 99.50 | 94.0 | 5.5 | -0.01 | 18.80 | -916 | -733 | 373 |
| 99.25 | 93.8 | 5.5 | -0.05 | 18.76 | -915 | -732 | 372 |
| 99.00 | 93.6 | 5.5 | -0.09 | 18.72 | -914 | -731 | 372 |
| 98.75 | 93.3 | 5.5 | -0.15 | 18.66 | -913 | -730 | 371 |
| 98.50 | 93.1 | 5.5 | -0.19 | 18.62 | -912 | -730 | 370 |
| 98.25 | 92.8 | 5.5 | -0.25 | 18.56 | -911 | -729 | 369 |
| 98.00 | 92.6 | 5.5 | -0.29 | 18.52 | -911 | -729 | 368 |

Case II:

Say the purchase of AN is for use in the manufacture of slurry and emulsion explosives. In this case generation of dirt and silica like materials for every 100 tons of Oxidizer solution prepared will be:

| WIM (%) | OB (ton) | AN (ton) | Insoluble per OB per 100 ton |
|---------|----------|----------|------------------------------|
| 0.05    | 15       | 0.525    | 52   kg                      |

This 52 kg, if not cleaned periodically from the oxidiser solution making tank, will affect the wear and tear of pumps in the plant.

## CONCLUSION

A long list of parameters of AN thus can be pruned into a small numbers of relevant parameters. The long form of specification was presented in the first section of this book. And the concise but still effective specification to accept consignments of AN can be pruned to the following:

| No. | PARAMETER | SPECIFICATION of AN for | |
|-----|-----------|-------------------------|------|
|     |           | Watergels and Emulsion | ANFO |
| 1 | Appearance | Without any visual impurities | Without any visual impurities |
| 2 | Physical form | HDAN | Free flowing, spherical prills |
| 3 | Purity | 99.5% Minimum, on dry basis | 98.0% Minimum, on dry basis |
| 4 | Shelflife of prills in the form of prills | 6 months | 6 months |
| 5 | Turbidity | Nil | Not Applicable |
| 6 | Bulk density | 1.00 to 1.05 g/cc | 0.70 to 0.80 g/cc |
| 7 | Water Insoluble Matter | 0.05% Maximum | 0.1% Maximum |

| | | | |
|---|---|---|---|
| 8 | Particle size<br>    More than 3 mm<br>    1 mm to 3 mm<br>    Less than 1 mm | 3% Maximum<br>95% Minimum<br>2% Maximum | Nil<br>98% Minimum<br>2% Maximum |
| 9 | Lab Batch | Gel/matrix should retain characteristics for 8 weeks in Hot Storage Test at 45 – 50°C | ANFO should absorb not more than 7% and retain not less than 5.5% oil |
| 10 | Oil Absorption | Not Applicable | 6.0 - 7 % |
| 11 | Oil Retention (up to 72 hours) | Not Applicable | 5.5 – 5.7 % |

# CHAPTER III

# CORELATIONS TABLES

Like the specification sheets, there are conversion tables that come handy for users of ammonium nitrate. Some of these are; (a) concentration determination from the crystallization point of the aqueous solutions of ammonium nitrate, (b) density concentration (c) AN and FO ratio table, and (d) ANFO SME ratio for the best output.

## Table I
## Density-Concentration Table

| Ammonium Nitrate ($NH_4NO_3$) in Water<br>https://handymath.com/cgi-bin/nh4no3tble.cgi?submit=Entry | | | | | | |
|---|---|---|---|---|---|---|
| Concentration (% Weight) | Temperature in degrees Centigrade (°C) | | | | | |
| | 0°C | 10°C | 25°C | 40°C | 60°C | 80°C |
| | Density (kg/L) | | | | | |
| 1 | 1.004 | 1.004 | 1.001 | 0.996 | 0.987 | 0.976 |
| 2 | 1.009 | 1.008 | 1.005 | 1.000 | 0.991 | 0.979 |
| 4 | 1.018 | 1.017 | 1.013 | 1.008 | 0.999 | 0.987 |
| 8 | 1.036 | 1.034 | 1.030 | 1.024 | 1.014 | 1.002 |
| 12 | 1.054 | 1.052 | 1.046 | 1.040 | 1.030 | 1.018 |
| 16 | 1.072 | 1.069 | 1.063 | 1.057 | 1.046 | 1.034 |
| 20 | 1.091 | 1.087 | 1.081 | 1.073 | 1.063 | 1.051 |
| 24 | 1.109 | 1.105 | 1.098 | 1.091 | 1.080 | 1.067 |
| 28 | 1.128 | 1.123 | 1.116 | 1.108 | 1.097 | 1.084 |
| 30 | 1.137 | 1.133 | 1.125 | 1.117 | 1.106 | 1.093 |
| 40 | 1.186 | 1.181 | 1.173 | 1.164 | 1.152 | 1.139 |
| 50 | 1.238 | 1.232 | 1.223 | 1.214 | 1.201 | 1.187 |

# Table II
# AN FO Ratio

| RATIO | | YIELD | | | | | |
|---|---|---|---|---|---|---|---|
| AN | FO | OB | SF | LF | AWS | ABS | VE |
| 94.00 | 6.00 | -1.77 | 18.80 | 6.0 | -969 | -775 | 377.0 |
| 94.10 | 5.90 | -1.41 | 18.82 | 5.9 | -958 | -766 | 376.6 |
| 94.20 | 5.80 | -1.05 | 18.84 | 5.8 | -948 | -758 | 376.1 |
| 94.30 | 5.70 | -0.69 | 18.86 | 5.7 | -938 | -750 | 375.7 |
| 94.40 | 5.60 | -0.32 | 18.88 | 5.6 | -928 | -742 | 375.2 |
| 94.45 | 5.55 | -0.14 | 18.89 | 5.6 | -923 | -738 | 375.0 |
| 94.46 | 5.54 | -0.10 | 18.89 | 5.5 | -922 | -738 | 374.9 |
| 94.47 | 5.53 | -0.07 | 18.89 | 5.5 | -920 | -736 | 374.9 |
| 94.48 | 5.52 | -0.03 | 18.90 | 5.5 | -919 | -735 | 374.8 |
| 94.49 | 5.51 | 0.00 | 18.89 | 5.5 | -918 | -734 | 374.8 |
| 94.50 | 5.5 | 0.04 | 18.86 | 5.7 | -917 | -734 | 374.8 |
| 94.51 | 5.4 | 0.40 | 18.52 | 7.4 | -916 | -733 | 374.7 |
| 94.52 | 5.3 | 0.77 | 18.17 | 9.1 | -915 | -732 | 374.7 |
| 94.53 | 5.2 | 1.13 | 17.83 | 10.8 | -914 | -731 | 374.6 |
| 94.54 | 5.1 | 1.49 | 17.49 | 12.6 | -913 | -730 | 374.6 |
| 95.00 | 5.0 | 1.85 | 17.15 | 14.3 | -866 | -693 | 372.5 |
| 95.10 | 4.9 | 2.22 | 16.80 | 16.0 | -856 | -685 | 372.1 |
| 95.20 | 4.8 | 2.58 | 16.46 | 17.7 | -846 | -677 | 371.6 |
| 95.30 | 4.7 | 2.94 | 16.12 | 19.4 | -835 | -668 | 371.2 |
| 95.40 | 4.6 | 3.31 | 15.77 | 21.1 | -825 | -660 | 370.7 |
| 95.50 | 4.5 | 3.67 | 15.43 | 22.8 | -815 | -652 | 370.3 |
| 95.60 | 4.4 | 4.03 | 15.09 | 24.6 | -804 | -643 | 369.8 |
| 95.70 | 4.3 | 4.40 | 14.74 | 26.3 | -794 | -635 | 369.4 |
| 95.80 | 4.2 | 4.76 | 14.40 | 28.0 | -784 | -627 | 368.9 |
| 95.90 | 4.1 | 5.12 | 14.06 | 29.7 | -774 | -619 | 368.5 |
| 96.00 | 4.0 | 5.48 | 13.72 | 31.4 | -763 | -610 | 368.0 |

AN: Ammonium Nitrate
FO: Fuel Oil
OB: Oxygen Balance
SF: Strength Factor
LF: Loss factor
ABS: Absolute Bulk Strength (Kcal/kg)
AWS: Absolute Weight Strength (Kcal/l)
VE: Volume Expansion

# Table III
# ANFO SME Ratio
# (This table is SME-specific)

| RATIO | | OUTPUT | | RATIO | | OUTPUT | |
| --- | --- | --- | --- | --- | --- | --- | --- |
| ANFO | SME | OB | SF | ANFO | SME | OB | SF |
| 0 | 100 | 4.00 | 0.00 | 50 | 50 | 0.50 | 10.00 |
| 2 | 98 | 3.86 | 0.40 | 52 | 48 | 0.36 | 10.40 |
| 4 | 96 | 3.72 | 0.80 | 54 | 46 | 0.22 | 10.80 |
| 6 | 94 | 3.58 | 1.20 | 56 | 44 | 0.08 | 11.20 |
| 8 | 92 | 3.44 | 1.60 | 58 | 42 | -0.06 | 11.60 |
| 10 | 90 | 3.30 | 2.00 | 60 | 40 | -0.20 | 12.00 |
| 12 | 88 | 3.16 | 2.40 | 62 | 38 | -0.34 | 12.40 |
| 14 | 86 | 3.02 | 2.80 | 64 | 36 | -0.48 | 12.80 |
| 16 | 84 | 2.88 | 3.20 | 66 | 34 | -0.62 | 13.20 |
| 18 | 82 | 2.74 | 3.60 | 68 | 32 | -0.76 | 13.60 |
| 20 | 80 | 2.60 | 4.00 | 70 | 30 | -0.90 | 14.00 |
| 22 | 78 | 2.46 | 4.40 | 72 | 28 | -1.04 | 14.40 |
| 24 | 76 | 2.32 | 4.80 | 74 | 26 | -1.18 | 14.80 |
| 26 | 74 | 2.18 | 5.20 | 76 | 24 | -1.32 | 15.20 |
| 28 | 72 | 2.04 | 5.60 | 78 | 22 | -1.46 | 15.60 |
| 30 | 70 | 1.90 | 6.00 | 80 | 20 | -1.60 | 16.00 |
| 32 | 68 | 1.76 | 6.40 | 82 | 18 | -1.74 | 16.40 |
| 34 | 66 | 1.62 | 6.80 | 84 | 16 | -1.88 | 16.80 |
| 36 | 64 | 1.48 | 7.20 | 86 | 14 | -2.02 | 17.20 |
| 38 | 62 | 1.34 | 7.60 | 88 | 12 | -2.16 | 17.60 |
| 40 | 60 | 1.20 | 8.00 | 90 | 10 | -2.30 | 18.00 |
| 42 | 58 | 1.06 | 8.40 | 92 | 8 | - | 18.40 |

| | | | |
|---|---|---|---|
| 44 | 56 | 0.92 | 8.80 |
| 46 | 54 | 0.78 | 9.20 |
| 48 | 52 | 0.64 | 9.60 |
| 50 | 50 | 0.50 | 10.00 |

| | | | |
|---|---|---|---|
| | | 2.44 | |
| 94 | 6 | -2.58 | 18.80 |
| 96 | 4 | -2.72 | 13.72 |
| 98 | 2 | -2.86 | 6.86 |
| 100 | 0 | -3.00 | 0.00 |

(in the above table, V = Volume Expansion)

The data on the cells of the above table is based on the AN and SME having following characteristics:

| PARAMETER | SME | ANFO |
|---|---|---|
| OB | 18.05 | 18.89 |
| BF | -0.06 | -0.04 |
| $\delta Hf$ | -623 | -917 |
| VE | 343 | 377 |

*-mkp-*

# REFERENCE

Allais, C. P., & Martinez, R. A. G. (2019). "Process for preparing a urea-sulphur fertilizer." U.S. Patent No. 10,464,855. Washington, DC: U.S. Patent and Trademark Office.

Albadarin A.B., Timothy D.L., Gavin M.W. Granulated polyhalite fertilizer caking propensity. Powder Technol. 2017;308:193.

Bachynsky J., Cheryl W., Krystal M. The practice of splitting tablets: Cost and therapeutic aspects. Pharmacoeconomics

Biamonte, Richard L., and Lorraine A. Corvino. Water soluble potassium phosphate caking inhibitor for fertilizer. U.S. Patent No. 5,286,272. 15 Feb. 1994.

Borman S. Advanced energetic materials emerge for military and space applications. Chem. Eng. News. 1994;72:18–22. doi: 10.1021/cen-v072n003.p018.

Chen T., Lv C. Study on Reducing AN Absorptivity with Surface Active Agents and Additives. Chin. J. Explos. Propellants. 1998;4:19–21.

Cui J., Han J., Wang J., Huang R. Study on the crystal structure and hygroscopicity of ammonium dinitramide. J. Chem. Eng. Data. 2010;55:3229–3234. doi: 10.1021/je100067n.

Clay R.B. Water-in-Oil Blasting Composition. 4,111,727. U.S. Patent. 1978 Sep 5;

Chaturvedi S., Dave P.N. Review on thermal decomposition of ammonium nitrate. J. Energetic Mater. 2013;31:1–26. doi: 10.1080/07370652.2011.573523.

Chattopadhyay A.K. Coating for Ammonium Nitrate Prills. 5,567,910. U.S. Patent. 1996 Oct 22;

Cohen J.S. Tablet splitting: imperfect perhaps, but better than excessive dosing. J. Am. Pharm. Assoc.

Chaturvedi S., Dave P. N., Review on thermal decomposition of ammonium nitrate, J. Energ. Mater. 31 (2013) (n.d.) 1.

Davey R., Ruddick A.J., Guy P.D., Mitchell B., Maginn S.J., Polywka L.A. The IV-III polymorphic phase transition in ammonium nitrate: A unique example of solvent mediation. J. Phys. D Appl. Phys. 1991;24:176. doi: 10.1088/0022-3727/24/2/014.

Daniel P. Barbis, Teddi S. Schaeffer; Titanium powders from the hydride–dehydride process; Titanium Powder Metallurgy, 2015

Eriz, Unai Elizundia, and Mateusz Marek Hass. Ammonium nitrate products and method for preparing the same. U.S. Patent Application No. 15/550,320.

Gezerman A.O. Effect of silicate, carbonate, calcium lignosulphonate, and silicic acid additives on degradation of ammonium nitrate. Kem Ind. 2020;3–4:20.

Gezerman A.O., Çorbacıoğlu B.D. Effects of sodium silicate, calcium carbonate, and silicic acid on ammonium nitrate degradation, and analytical investigations of the degradation process on an industrial scale. Chem. Ind. Chem. Eng. Q. 2015;21:359.

Gezerman A.O., Corbacioglu B.D. Caking and degradation problem on nitrogenous fertiliser and alternative solution processes. Int. J. Chem. 2011;3:123.

Gezerman A.O., Çorbacioglu B.D. Detonation properties of ammonium nitrate containing calcium carbonate, dolomite, and fly ash. Int. J. Energ. Mater. Chem. Propuls. 2017;16:295.

Gezerman A.O. Istanbul, Yildiz Technical University, Dissertation for the Degree of Doctor Philosophiae; 2018. Alternative Methods to Prevent of Caking and Degradation of Nitrogenous Fertilizers Used in Turkey.

Gezerman A.O., Corbacioglu B.D., Cevik H. Improvement of surface features of nitrogenous fertilisers and influence of surfactant composition on fertiliser surface. Int. J. Chem. 2011;3:201.

Gang Y.P., Cang S.W., Zhang Y.H., Li Q.S. Self assembling Monolayers for Anti-moisture Absorption of Ammonium Nitrate. Chin. J. Appl. Chem. 2000;17:186–188.
Gunawan R., Freij S., Zhang D.K., Beach F., Littlefair M. A mechanistic study into the reactions of ammonium nitrate with pyrite. Chem. Eng. Sci. 2006;61:5781–5790. doi: 10.1016/j.ces.2006.04.044.

https://pharmaceuticalupdates.com/2019/02/17/table t-friability-test-specification-and-calibration/

https://www.sciencedirect.com/topics/chemistry/friabil ity
http://www.fertilizerseurope.com/fileadmin/user_uploa d/user_upload_prodstew/documents/Booklet_nr_6_Production_ of_Ammonium_Nitrate_and_Calcium_Ammonium_Nitrate.pdf

Heintz T., Pontius H., Aniol J., Birke C., Leisinger K., Reinhard W. Ammonium Dinitramide (ADN)—Prilling, Coating, and Characterization. Propellants Explos. Pyrotech. 2009;34:231–238. doi: 10.1002/prep.200800110.

Harris J. Hygroscopicity of Ammonium Nitrate Samples. DTIC Document; Fort Belvoir, VA, USA: 1970.

Heller A.N., Cuffe S.T., Goodwin D.R. Sources of Air Pollution and Their Control: Air Pollution. Academic Press; Cambridge, MA, USA: 2015. Inorganic chemical industry; p. 191.

Han Z. Texas A&M University; PhD Thesis: 2016. Thermal Stability Studies of Ammonium Nitrate.

Hay, D.N., Dimas, P.A., 2018.(Ecolab USA Inc.), U.S. Patent No. 9,353,301 B2. Hay, Daniel NT, and Peter A. Dimas. Composition for dust control. U.S. Patent No. 9,353,301. 31 May 2016.

Huang W.Y., Yan S.L., Xie X.H., Guo Z. The production and Implication of Composite Modified Water-Resistant Powdery

Ammonium Nitrate; Proceedings of the International Conference on Theory and Practice of Energetic Materials (IASPEP 2003); Guilin, China. 15–18 October 2003; pp. 68–73.

Damse R. Waterproofing materials for ammonium nitrate. Def. Sci. J. 2004;54:483–492. doi: 10.14429/dsj.54.2062.

Hu K.L., Luo N., Huang W.Y., Ma X.M., Xu X.F. Test and Study on the Modification of Ammonium N itrate by Coating Its Surface. Explos. Mater. 2006;35:14–17.

Jablon, Michael, Maria Anatolyevna Azimova, and Gerald Smith. Wax-based fertilizer coatings with polyethylene-or polypropylene-based polymers. U.S. Patent No. 10,081,578. 25 Sep. 2018.

J. Skopp; PARTICLE SIZE SEPARATION | Sieving/Screening; Encyclopedia of Separation Science, 2000

Joshi A.A., Duriez X. Added functionality excipients: an answer to challenging formulations. Pharm. Technol. Suppl

Komunjer L., Affolter C. Absorption-evaporation kinetics of water vapour on highly hygroscopic powder: Case of ammonium nitrate. Powder Technol. 2005;157:67–71. doi: 10.1016/j.powtec.2005.05.012.

Komunjer L., Pezron I. A new experimental method for determination of solubility and hyper-solubility of hygroscopic solid. Powder Technol. 2009;190:75–78. doi: 10.1016/j.powtec.2008.04.060.

Kristensen H.G., Jorgensen G.H., Sonnergaard J.M. Mass uniformity of tablets broken by hand. Pharmeuropa

Kim J.-K., Choi S.-I., Kim E.J., Kim J.H., Koo K.-K. Preparation of Spherical Ammonium Nitrate Particles by Melt Spray. Ind. Eng. Chem. Res. 2010;49:12632–12637. doi: 10.1021/ie1006992.

Kajiyama K., Izato Y.-I., Miyake A. Thermal characteristics of ammonium nitrate, carbon, and copper (II) oxide mixtures. J. Therm. Anal. Calorim. 2013;113:1475–1480. doi: 10.1007/s10973-013-3201-5.

Kwok Q.S., Kruus P., Jones D.E. Wettability of ammonium nitrate prills. J. Energetic Mater. 2004;22:127–150. doi: 10.1080/07370650490522776.

Lu L.Y., Yang L., Zhou Y.K. Study on Contact Angle for Characterizing the Modifying Effect of Powder Ammonium Nitrate. Coal Mine Blasting. 2009;4:4–6.

Lu L.-Y., Yang L. Study on the Interaction Mechanism of Modified Ammonium Nitrate by O tadecylamine. Initiat. Pyrotech. 2010;2:41–43.

Marlair G., Kordek M.-A. Safety and security issues relating to low capacity storage of AN-based fertilizers. J. Hazard. Mater. 2005;123:13–28. doi: 10.1016/j.jhazmat.2005.03.028.

Malash G., Hashem H. Improving the properties of ammonium nitrate fertilizer using additives. Alex. Eng. J. 2005;44:685–693.

Masao K., Asaji K., Noriynki M., Tsutomu Y. Encapsulation Method. 3,691,090. U.S. Patent. 2014 Octomber;

Mellor J.W. A Comprehensive Treatise on Inorganic and Theoretical Chemistry. Volume 2 Longmans, Green; London, UK: 1922.

Martinez, Joan Antoni Riaza, and Marc Rocafull Fajardo. Anti-caking compositions for fertilizers. U.S. Patent No. 8,932,490. 13 Jan. 2015.

M. N. Sudin, S. Ahmed-Kristensen and M. M. Andreasen; The role of a specification in the design process; https://www.researchgate.net/publication/265158743

Martinez, Joan Antoni Riaza, and Marc Rocafull Fajardo. Anti-caking compositions for fertilizers. U.S. Patent No. 8,932,490. 13 Jan. 2015.

Matsunaga H., Habu H., Miyake A. Thermal decomposition of the high-performance oxidizer ammonium dinitramide under pressure. J. Therm. Anal. Calorim. 2014;116:1227–1232. doi: 10.1007/s10973-013-3626-x.

Mathew S., Krishnan K., Ninan K. Effect of energetic materials on thermal decomposition of phase-stabilised ammonium nitrate-An eco-friendly oxidiser. Def. Sci. J. 1999;49:65–69. doi: 10.14429/dsj.49.3788.

Patrick F.D., Rapstein K.T., Schrieber C.F., Wilson J.S. Coated Explosive Comprising Ammonium Nitrate. 3,287,189. U.S. Patent. 1966 Nov 22;

N.D. Oltargevskaya, G.E. Krichevsky; CHOICE OF TEXTILE MATERIAL, TECHNOLOGY, POLYMER – THICKENER AND MEDICAL PRODUCT; Medical Textiles and Biomaterials for Healthcare, 2006

Nidal H. Daraghmeh, Adnan A. Badwan; Profiles of Drug Substances, Excipients and Related Methodology, 2011

Ogzewalla, Mark B., Archimedo Mario Carlini Jr, and James J. Barnat. Dust and anticaking resistant fertilizer. U.S. Patent Application No. 15/404,348.

Obrestad, Torstein, and Terje Tande. Conditioning agent for a particulate fertilizer for reducing hygroscopicity and dust formation. U.S. Patent No. 10,294,170. 21 May 2019.

Oommen C., Jain S. Ammonium nitrate: A promising rocket propellant oxidizer. J. Hazard. Mater. 1999;67:253–281. doi: 10.1016/S0304-3894(99)00039-4.

Oxley J.C., Smith J.L., Rogers E., Yu M. Ammonium nitrate: Thermal stability and explosivity modifiers. Thermochim. Acta. 2002;384:23–45. doi: 10.1016/S0040-6031(01)00775-4.

Parker, D. Todd, Mark Ogzewalla, and Zachary T. Burrell. Fertilizer coating for dust control and/or anti-caking. U.S. Patent Application No. 16/126,334.

R.A. Shanks; Alternative solutions: recyclable synthetic fibre–thermoplastic composites; Green Composites, 2004

Rodenhuis N., De Smet P.A.G.M., Barends D.M. Th e rationale of scored tablets as dosage form. Eur. J. Pharm. Sci

Rosen M.J., Kunjappu J.T. Surfactants and Interfacial Phenomena. John Wiley & Sons; Hoboken, NJ, USA: 2012.

Myers D. Surfactant Science and Technology. John Wiley & Sons; Hoboken, NJ, USA: 2005.

Sara Nilsson, and MattiasLindahl; A Literature Review to Understand the Requirements Specification's Role when Developing Integrated Product Service Offerings; https://www.sciencedirect.com/science/article/pii/S22128271 16304504

Savci S. An agricultural pollutant: chemical fertilizer. Int. J. Environ. Sci. Technol. 2012;3:73

Strydom C.A., Mthombo T.S., Bunt J.R., Neomagus H.W.J.P. Some physical and chemical characteristics of calcium lignosulphonate-bound coal fines. J. S. Afr. Inst. Min. Metall. 2018; 118:1277–1283.

Sinditskii V.P., Egorshev V.Y., Levshenkov A.I., Serushkin V.V. Ammonium nitrate: Combustion mechanism and the role of additives. Propellants Explos. Pyrotech. 2005;30:269–280. doi: 10.1002/prep.200500017.

Singh G., Felix S. Evaluation of transition metal salts of NTO as burning rate modifier for HRPB-AN composite solid propellants. Combust. Flame. 2003;132:422–432. doi: 10.1016/S0010-2180(02)00479-0.

Trevor F. Starr; Pultrusion applications – a world-wide review; Pultrusion for Engineers, 2000

Umer H., Nigam H., Tamboli A.M., Nainar M.S.M. Microencapsulation: Process, techniques and applications. Int. J. Res. Pharm. Biomed. Sci. 2011;2:474–480.

Ushijima, Helio Haruo, and Ricardo Chagas da Silva. Concentrated sugar additive as anti-dusting agent. U.S. Patent No. 9,708,520. 18 Jul. 2017.

Varshovi, Allan Amir. Coated organic materials and methods for forming the coated organic materials. U.S. Patent No. 10,384,984. 20 Aug. 2019.

Vargeese A.A., Muralidharan K., Krishnamurthy V. Thermal stability of habit modified ammonium nitrate: Insights from isoconversional kinetic analysis. Thermochim. Acta. 2011;524:165–169. doi: 10.1016/j.tca.2011.07.009.

Van Santen E., Barends D.M., Frijlink H.W. Breaking of scored tablets: a review. Eur. J. Pharm. Biopharm.

Vranić E., Uzunović A. Infl uence of tablet splitting on content uniformity of lisinopril/ hydrochlorthiazide tablets. Bosn. J. Basic Med. Sci.

Van Nieuwenhuyse A.E. European Fertilizer Manufacturers' Association; Brussels: 2000. Production of Ammonium Nitrate and Calcium Ammonium Nitrate in Booklet No.

Wei Y., Cai B.H. Study on surface modification of ammonium nitrate. Adv. Mater. Res. 2012; 399:1989.

Wei Y., Cai B.H. Advanced Materials Research. Trans Tech Publications; Pfaffikon, Switzerland: 2011. Study on Surface Modification of Ammonium Nitrate.

Wen Y. Master's Thesis. Hunan University; Changsha, China: 2003. Study on The Technics of Surface Modification of Ammonium Nitrate.

Xiong Y.-T., Liu Z.-L. Improvement of the Hygroscopicity of AN by Modified Paraffin. Chin. J. Explos. Propellants. 2013;4:50–53.

Xu J.-M., Liu Z.-L., Zhai D.D. Modification of inorganic salts by encapsulation method. Liaoning Chem. Ind. 2003;32:142.

Ye Z.-W. Research on Improvement in Hygroscopicity of Ammonium Nitrate with Surfactants. Fine Chem. Dalian. 2001;18:70–71.

Ye W. Research on properties of ammonium nitrate modified by coupling cationic surfactant. Inorg. Chem. Ind. 2007;39:26.

Yu Z.C., Zhang Y.J.; Master's Thesis. Nanjing University of Science & Technology; Nanjing, China: 2013. Coating and modification of ammonium nitrate particles.

Yin P.-G., Jin G.-L., Zhang Y.-H., Li Q.-S. Structure Characterization of Self-Assembled Monolayer Film on the Surface of Ammonium Nitrate. J. Beijing Inst. Technol. 2000;20:643–646.

Yin Y.-X., Yang R.-J. Modification of AN by using microencapsulation technology. Liaoning Chem. Ind. 2001;30:139. doi: 10.3969/j.issn.1004-0935.2001.04.001.

Zhang X.D., Li J.M., Yang R.J., Zhao X.Q. Surface Modification of Phase Stabilized Ammonium Nitrate and Its Application in Solid Composite Propellants. Chin. J. Explos. Propellants. 2009;32:5–9.

Zhang J. Study on Properties of the Coated AN with Polyvinyl Butyral. Chin. J. Explos. Propellants. 2001;24:41–43.

Zurimendi, Jon. Anti-caking composition. U.S. Patent No. 4,772,308. 20 Sep. 1988.

Zhang J., Yang R.-J. Study on Surface Properties of Coated Ammonium Nitrate. Energetic Mater. 2004;12:1–5.

Zhang J.Z., Wang X.Q., Lin D.-H. Study on Hygroscopicity of Ammonium Nitrate Particle Coated by Precipitation Polymerization of Styrene. Adv. Fine Petrochem. 2008;12:35–37.

*-mkp-*